OH YEAH! OH MY! LET'S MULTIPLY!

Kenya Taylor-Wash, M.Ed.

Welcome

Hi there!

Did you know that learning your multiplication facts can make math easier for you? Well, it's true. It can even make math fun, unless you're like me and already know that math is fun!

This guide will help you learn your multiplication facts from 0 to 12. If you study every day, you will be multiplying in no time. Then you will be able to pass this study guide on to someone else. Study well, and have fun!

Sincerely,

Kenya Taylor-Wash, M.Ed.

Study Tips

- Study the key words in the glossary.
- Study one number at a time.
- Ask a helper to quiz you.
- Take notes.
- Practice with flashcards.
- Know the Commutative Property of Multiplication!

The Commutative Property of Multiplication:

changing the order of factors does not change the product. Example: 3 x 5 is the same as 5 x 3

They both equal 15!

ZERO

Classification:
- whole number
- neutral integer

~~odd~~ or **even**

0

~~prime or composite~~
*NEITHER

*The multiples of 0 are ALL 0!

Practice Multiplying by 0

STUDY TIP:
Any number x 0 = 0!

0 x 1 = 0
0 x 2 = 0
0 x 3 = 0
0 x 4 = 0
0 x 5 = 0
0 x 6 = 0
0 x 7 = 0
0 x 8 = 0
0 x 9 = 0
0 x 10 = 0
0 x 11 = 0
0 x 12 = 0

The commutative Property

3 x 0 = 0 x 3

ONE

Classification:
- whole number
- positive integer

1

odd or ~~even~~

~~prime or composite~~
NEITHER
*only 1 factor: 1 x 1 = 1

The First 12 Multiples of 1

1
2
3
4
5
6
7
8
9
10
11
12

Multiplying by 1

1 x 0 = 0
1 x 1 = 1
1 x 2 = 2
1 x 3 = 3
1 x 4 = 4
1 x 5 = 5
1 x 6 = 6
1 x 7 = 7
1 x 8 = 8
1 x 9 = 9
1 x 10 = 10
1 x 11 = 11
1 x 12 = 12

STUDY TIP:
Any number x 1 = that number!

The Commutative Property

1 x 4 = 4 x 1

Practice Multiplying by 1

1 x 0 = ?

1 x 1 = ?

1 x 2 = ?

3 x 1 = ?

1 x 4 = ?

5 x 1 = ?

1 x 6 = ?

7 x 1 = ?

1 x 8 = ?

9 x 1 = ?

1 x 10 = ?

11 x 1 = ?

1 x 12 = ?

Speed alert! Go fast when you see gray. You studied it already!

TWO

Classification:

- whole number
- positive integer

2

~~odd~~ or **even**

prime or ~~composite~~

*only two factors: 1 x 2 = 2

The First 12 Multiples of 2

PATTERN ALERT!
The Multiples of 2 are ALL even!

2
4
6
8
10
12
14
16
18
20
22
24

Multiplying by 2

STUDY TIP:
When multiplying by 2, double the other number!

$2 \times 8 = 16$
$8 + 8 = 16$

2 x 0 = 0
2 x 1 = 2
2 x 2 = 4
2 x 3 = 6
2 x 4 = 8
2 x 5 = 10
2 x 6 = 12
2 x 7 = 14
2 x 8 = 16
2 x 9 = 18
2 x 10 = 20
2 x 11 = 22
2 x 12 = 24

Practice Multiplying by 2

0 x 2 = ?
2 x 1 = ?
2 x 2 = ?
2 x 3 = ?
4 x 2 = ?
2 x 5 = ?
6 x 2 = ?
2 x 7 = ?
8 x 2 = ?
2 x 9 = ?
10 x 2 = ?
2 x 11 = ?
12 x 2 = ?

Speed alert!
You studied these already!

THREE

Classification:
- whole number
- positive integer

3

odd or ~~even~~

prime or ~~composite~~

*only two factors: 1 x 3 = 3

The First 12 Multiples of 3

3
6
9
12
15
18
21
24
27
30
33
36

Multiplying by 3

3 x 0 = 0

3 x 1 = 3

3 x 2 = 6

3 x 3 = 9

3 x 4 = 12

3 x 5 = 15

3 x 6 = 18

3 x 7 = 21

3 x 8 = 24

3 x 9 = 27

3 x 10 = 30

3 x 11 = 33

3 x 12 = 36

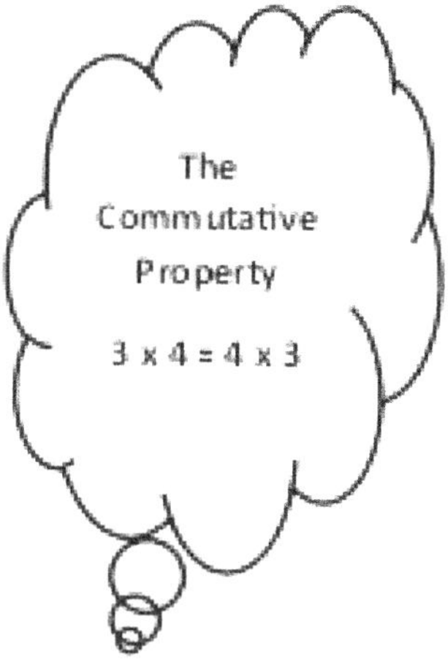

The First 12 Multiples of 4

4
8
12
16
20
24
28
32
36
40
44
48

Multiplying by 4

4 x 0 = 0
4 x 1 = 4
4 x 2 = 8
4 x 3 = 12
4 x 4 = 16
4 x 5 = 20
4 x 6 = 24
4 x 7 = 28
4 x 8 = 32
4 x 9 = 36
4 x 10 = 40
4 x 11 = 44
4 x 12 = 48

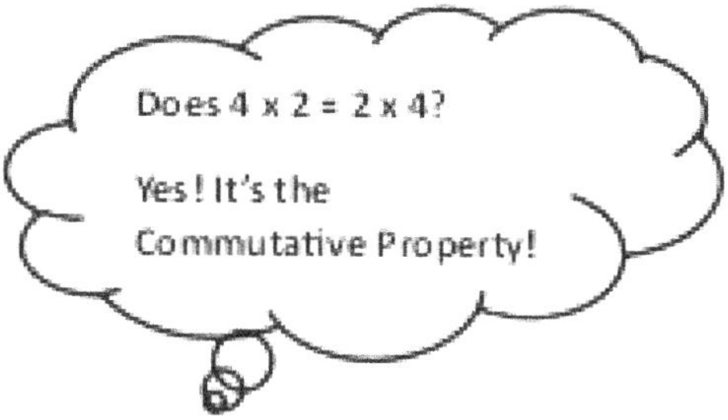

Practice Multiplying by 4

0 x 4 = ?
4 x 1 = ?
2 x 4 = ?
4 x 3 = ?
4 x 4 = ?
4 x 5 = ?
6 x 4 = ?
4 x 7 = ?
8 x 4 = ?
4 x 9 = ?
10 x 4 = ?
4 x 11 = ?
12 x 4 = ?

FIVE

Classification:
- whole number
- positive integer

5

odd or ~~even~~

prime or ~~composite~~

*only two factors: 1 x 5 = 5

Practice Multiplying by 5

5 x 0 = ?
1 x 5 = ?
5 x 2 = ?
3 x 5 = ?
5 x 4 = ?
5 x 5 = ?
5 x 6 = ?
7 x 5 = ?
5 x 8 = ?
9 x 5 = ?
5 x 10 = ?
11 x 5 = ?
5 x 12 = ?

Practice Multiplying by 5

5 x 0 = ?
1 x 5 = ?
5 x 2 = ?
3 x 5 = ?
5 x 4 = ?
5 x 5 = ?
5 x 6 = ?
7 x 5 = ?
5 x 8 = ?
9 x 5 = ?
5 x 10 = ?
11 x 5 = ?
5 x 12 = ?

Multiplying by 5

5 x 0 = 0

5 x 1 = 5

5 x 2 = 10

5 x 3 = 15

5 x 4 = 20

5 x 5 = 25

5 x 6 = 30

5 x 7 = 35

5 x 8 = 40

5 x 9 = 45

5 x 10 = 50

5 x 11 = 55

5 x 12 = 60

STUDY TIP:
When multiplying by 5, the product always ends in 0 or 5!

LET'S GET NERDY:
When multiplying by 5, if the other factor is even, the product will end in zero but if the other factor is odd, the product will end in 5!

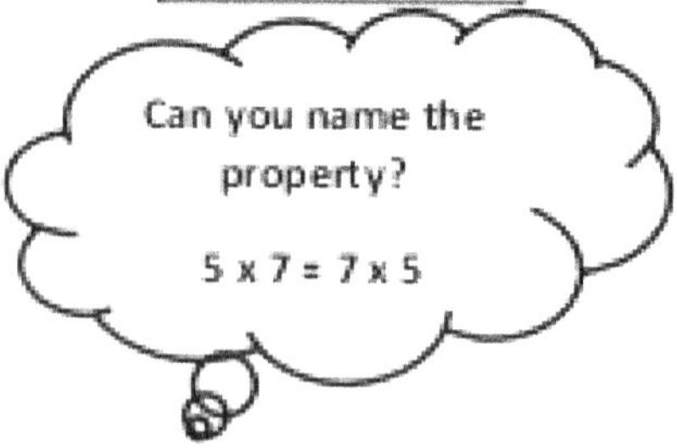

The First 12 Multiples of 5

PATTERN ALERT!
The Multiples of 5 end in 0 or 5!

5
10
15
20
25
30
35
40
45
50
55
60

FIVE

Classification:
- whole number
- positive integer

5

odd or ~~even~~

prime or ~~composite~~

*only two factors: 1 x 5 = 5

SIX

Classification:
- whole number
- positive integer

6

~~odd~~ or **even**

~~prime~~ or **composite**

*more than 2 factors!

1 x 6 = 6

2 x 3 = 6

The factors of 6: 1, 2, 3, 6

The First 12 Multiples of 6

6
12
18
24
30
36
42
48
54
60
66
72

Multiplying by 6

6 x 0 = 0
6 x 1 = 6
6 x 2 = 12
6 x 3 = 18
6 x 4 = 24
6 x 5 = 30
6 x 6 = 36
6 x 7 = 42
6 x 8 = 48
6 x 9 = 54
6 x 10 = 60
6 x 11 = 66
6 x 12 = 72

The First 12 Multiples of 7

7
14
21
28
35
42
49
56
63
70
77
84

The First 12 Multiples of 7

7

14

21

28

35

42

49

56

63

70

77

84

ing by 7

= 0

= 7

= 14

= 21

= 28

= 35

= 42

= 49

= 56

= 63

0 = 70

1 = 77

2 = 84

Practice Multiplying by 7

7 x 0 = ?
1 x 7 = ?
7 x 2 = ?
3 x 7 = ?
7 x 4 = ?
5 x 7 = ?
7 x 6 = ?
7 x 7 = ?
7 x 8 = ?
9 x 7 = ?
7 x 10 = ?
11 x 7 = ?
7 x 12 = ?

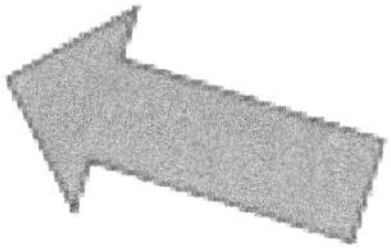

Speed alert!

You've come a long way!

ng by 6

= 0
= 6
= 12
= 18
= 24
= 30
= 36
= 42
= 48
= 54
= 60
= 66
= 72

EIGHT

Classification:
- whole number
- positive integer

8

~~odd~~ or **even**

~~prime~~ or **composite**

*more than 2 factors!

1 x 8 = 8

2 x 4 = 8

The factors of 8: 1, 2, 4, 8

The First 12 Multiples of 8

8
16
24
32
40
48
56
64
72
80
88
96

Multiplying by 8

8 x 0 = 0
8 x 1 = 8
8 x 2 = 16
8 x 3 = 24
8 x 4 = 32
8 x 5 = 40
8 x 6 = 48
8 x 7 = 56
8 x 8 = 64
8 x 9 = 72
8 x 10 = 80
8 x 11 = 88
8 x 12 = 96

Practice Multiplying by 8

0 x 8 = ?
8 x 1 = ?
2 x 8 = ?
8 x 3 = ?
4 x 8 = ?
8 x 5 = ?
6 x 8 = ?
8 x 7 = ?
8 x 8 = ?
8 x 9 = ?
10 x 8 = ?
8 x 11 = ?
12 x 8 = ?

NINE

Classification:

- Odd composite whole number
- positive integer

9

LET'S GET NERDY:
9 is the only **odd** number less than 12 that's also **composite**!

odd or ~~even~~

~~prime~~ or **composite**

*more than 2 factors!
<u>1</u> x <u>9</u> = 9
<u>3</u> x 3 = 9
The factors of 9: 1, 3, 9

The First 12 Multiples of 9

9
18
27
36
45
54
63
72
81
90
99
108

Multiplying by 9

9 x 0 = 0
9 x 1 = 9
9 x 2 = 18
9 x 3 = 27
9 x 4 = 36
9 x 5 = 45
9 x 6 = 54
9 x 7 = 63
9 x 8 = 72
9 x 9 = 81
9 x 10 = 90
9 x 11 = 99
9 x 12 = 108

The First 12 Multiples of 10

PATTERN ALERT!
The Multiples of 10 ALL end in 0!

10
20
30
40
50
60
70
80
90
100
110
120

Multiplying by 10

10 x 0 = 0
10 x 1 = 10
10 x 2 = 20
10 x 3 = 30
10 x 4 = 40
10 x 5 = 50
10 x 6 = 60
10 x 7 = 70
10 x 8 = 80
10 x 9 = 90
10 x 10 = 100
10 x 11 = 110
10 x 12 = 120

Practice Multiplying by 10

0 x 10 = ?
10 x 1 = ?
2 x 10 = ?
10 x 3 = ?
4 x 10 = ?
10 x 5 = ?
6 x 10 = ?
10 x 7 = ?
8 x 10 = ?
10 x 9 = ?
10 x 10 = ?
10 x 11 = ?
12 x 10 = ?

ELEVEN

odd or ~~even~~

11

prime or ~~composite~~

*only two factors: 1 x 11 = 11

Practice Multiplying by 11

11 x 0 = ?
1 x 11 = ?
11 x 2 = ?
3 x 11 = ?
11 x 4 = ?
5 x 11 = ?
11 x 6 = ?
7 x 11 = ?
11 x 8 = ?
9 x 11 = ?
11 x 10 = ?
11 x 11 = ?
11 x 12 = ?

Ke

Practice Multiplying by 11

11 x 0 = ?
1 x 11 = ?
11 x 2 = ?
3 x 11 = ?
11 x 4 = ?
5 x 11 = ?
11 x 6 = ?
7 x 11 = ?
11 x 8 = ?
9 x 11 = ?
11 x 10 = ?
11 x 11 = ?
11 x 12 = ?

ELEVEN

odd or ~~even~~

11

prime or ~~composite~~

*only two factors: <u>1</u> x <u>11</u> = 11

TWELVE

~~odd~~ or even

12

~~prime~~ or **composite**

*more than 2 factors!
1 x 12 = 12
2 x 6 = 12
3 x 4 = 12
The factors of 12: 1, 2, 3, 4, 6, 12

The First 12 Multiples of 12

12
24
36
48
60
72
84
96
108
120
132
144

Multiplying by 12

12 x 0 = 0
12 x 1 = 12
12 x 2 = 24
12 x 3 = 36
12 x 4 = 48
12 x 5 = 60
12 x 6 = 72
12 x 7 = 84
12 x 8 = 96
12 x 9 = 108
12 x 10 = 120
12 x 11 = 132
12 x 12 = 144

Factor - a number that is multiplied by another number

EXAMPLE:

2 x 3 = 6

2 and 3 are factors of 6

Integers – all numbers less than 0, the number zero, and all numbers greater than 0 (no fractions or decimals)

EXAMPLE:

... -5, -4, -3 -2, -1, 0, 1, 2, 3, 4, 5, ...

Multiples – the numbers that are made after multiplying a given whole number by other whole numbers

EXAMPLE:

3 x 1 = 3

3 x 2 = 6

3 x 3 = 9

3, 6, 9 are the first three multiples of 3

Odd numbers - numbers that cannot be divided into two equal groups.

EXAMPLE: 1, 3, 5, 7, 9, 11 are odd numbers

Prime numbers – numbers that have only 2 factors, 1 and the number

EXAMPLE:

11 is prime
*only two factors: 1 x 11 = 11
The factors of 11: 1 and 11

LET'S GET NERDY:
ALL **prime** numbers are **odd**
except the number 2!

Product – the answer to a multiplication problem

EXAMPLE:

3 x 4 = 12 (12 is the product)

Whole numbers – all numbers starting from zero (they do not include fractions or decimals)

EXAMPLE: 0, 1, 2, 3, 4, 5, and so forth

NOTES

NOTES

Made in the USA
Monee, IL
04 November 2026

67386524R00036